Yes, No, Maybe, Computes

Filters That Can Increase Gain On Computing

Edward M Seymour

This volume is dedicated to those curious to see a way
computing can change to become more intuitive.

Table of Contents

About the author

I began my introduction to computing in high school so long ago the galaxy far away was not even devised. It was the age of Aquarius where freedom to experiment was rampant and wars were a thing that divided us from the establishment. Computing was sitting at a terminal remote connected via a telephone to a box at Syracuse University. This was a rare privilege to have such access and ability to "program" in Basic, Fortran or Apl. Programs were saved on stacks of cards, each having 8 characters for comment or number and 72 characters for code.

During the same era books like 1984 by George Orwell and Soylent Green by Harry Harrison were required reading. Concerns over Big Brother and associated government control dominated the landscape. This for me was my launch into the world of computing. Since then I have worked on computer development in IBM and subsequent places that have allowed me to work on processors that range from Unix servers to military controllers and included a period of helping develop curriculum for undergraduate study of chip design and test along with supporting the Auto Industry and compute intense applications.

The ideas contained in this book encapsulate observations I have developed over decades of computer development and observation of how people make decisions. These concepts revealed in this book are soley mine and are not affiliated with any past or present employer. I do firmly believe, however that applications of these concepts can help futher the advancement of computing as we know it. In many way you could say I offer filters for big data and cloud computing that will help harvest more of the existing potential.

Acknowledgements

My philosophy of how computing could more closely match our expectations has come from decades of study. Study, in the strict sense has been an amazing split of classroom, laboratory merged with an art form called people watching. One could say it is the last element where I have gained the most. In essence it is as though I have been an apprentice in several disciplines. It has been a form of "learn from others mistakes, successes, break evens". While the list of others is literally much too long to reasonably include, I must mention with great appreciation the encouragement and perspective I have gained from a select few people.

My first thanks go to my supportive wife and family without whose support, none of this would be possible. In the realm of family a special tribute goes to my Mom, Doris Seymour. While she left us in 2004, she left behind such a legacy. She was, first and foremost, an acute observer of human behavior, where her means for recording this was poetry. For as long as she was alive she wrote 2-3 poems a day and handed them out to random people crossing here path. Upon her death, she handed me that baton. A chronicle of her amazing life will be the subject of another volume.

Next up, my Dad, Mason Seymour, taught me that compassion and love were best expressed in actions. Words for him were measured and carefully dispensed but actions speak louder than words rang so true. His capacity for love and compassion was unlimited, even in the face of adversity. He was a strong believer in Fate and the idea that each of us was on a path, predetermined.

Enter my Mom's mother, maternal grandmother, Sarah Southworth, who had survived a harsh life. For her the depression was a painful experience where she lost a daughter while the daughter was giving birth, gave her only son to tragedy in World War II, nearly lost my mom to sickness and was hopelessly addicted to caffeine. She was amazingly unflappable but held a hard line on one concept, never tell her that you can't do something. The result is, out comes the lecture.

Jack Mosher came next as imitation Grandfather. This was due to the following factors. All other biological grandparents had died before I was born. In addition, Jack had this innate curiosity and passion for life. His contribution to this collection was the value of using correct, simple tools whenever you can. His ability to pull one's gullible leg and use humor of catastrophe has stuck with me to this day.

Andre Lepine came into our lives as a French Canadian bushman who taught Agriculture at my school. From him I gained immense respect for mother nature, ability to identify plants and trees along with some of the most novel ways to employ them. To his credit I got immersed in French. To his credit, he had a devious way to coerce teenagers to gain respect for their own ability and worth. He was also a miracle worker with rough, unsawn wood.

Rochester Institute of Technology

http://www.rit.edu/ was where I learned that the study of math and science form a gateway through which one could apply critical thinking skills. Firmly embedded in the DNA of the place was the insistence on practicing the craft of which you study. In this test drive of a career path, you either affirm or deny life long interests. It was a private university with vocational focus and demand of excellence.

Dr Kenneth Hsu (https://people.rit.edu/kwheec/homepage.html) was a passionate junior professor when he joined our fray at RIT. In time, it became clear that what he lacked in English skills was completely compensated by his depth of knowledge in computer architecture and computer systems. As his first class of cynical students, we were unkind and unforgiving. We even developed a hash table to translate a language we phrased as Ah Chu into English.

Dr Roy Czernikowski

https://www.linkedin.com/in/roy-czernikowski-3aa0ba7a?authType=NAME_SEARCH&authToken=IDq2&locale=en_US&trk=tyah&trkInfo=clickedVertical%3Amynetwork%2CclickedEntityId%3A280928434%2CauthType%3ANAME_SEARCH%2Cidx%3A1-2-2%2CtarId%3A1473781947738%2Ctas%3ARoy%C2%A0Czernikowski

He was the mad man behind a revolution called Computer Engineering. His method of validating that you listened in class was Baptism by fire followed by a walk on hot coals. His passion was real time computing and it came with a focus on what you could do with 100 lines of code or less. This included programming to avoid train collisions while maximizing the speed of trains on a collision course.

Dr Roger Heinz

Dr Heinz was a large, imposing German genius who demanded deep comprehension of electronics that was co-requisite with use of higher level math. His job as instructor seemed to be two fold. Put the wrath of gods in your mind. Then listen, observe and obtain the core idea by means of visualizing a solution via simple graphs. Know your derivations was a bonus when bored.

Penn State University

http://www.psu.edu/

Dr David Landis – (now at CMU instead)

https://www.cmu.edu/engineering/materials/people/faculty/bios/landis.html

Dave was a partner in a ground breaking effort to prove it possible to teach computer architecture, design and test to undergraduate students in the mid 1980s. (https://www.linkedin.com/in/edwardmseymour?trk=hp-identity-name) – [see publications] While student chip samples produced and validated were relatively small, it was unpredicted to see this play out in a single semester class.

University Of Illinois, Urbana–Champaign

http://illinois.edu/

Dr William Kubitz

https://cs.illinois.edu/directory/profile/kubitz

Dr Kubitz was another parter in the aforementioned program who had students creating silicon targeted at hardware implementation of software algorithms. Strictly speaking Dr Kubitz was in Computer Science department doing engineering stuff.

Dr Janak Patel

https://www.ece.illinois.edu/directory/profile/jhpatel

Dr Patel had students who tested out computer architecture in silicon. His efforts in the same program added yet another element of depth in the study of computers.

Dr Jacob Abraham

http://www.ece.utexas.edu/people/faculty/jacob-abraham

Dr Abraham was at University of Illinois when I had the pleasure to collaborate. His students were pushing new architecture thoughts in silicon as well.

University of Minnesota

https://twin-cities.umn.edu/

Dr Gerald Sobelman

http://mountains.ece.umn.edu/~sobelman/

Dr Sobelman came to the program a bit later than the others in that Minnesota was selected forth. He ha come from industry to become a professor and thus brought a different, more vocational aspect. His students had to run a gauntlet of qualifying runs before being selected for fabrication.

University of Tennesee

https://www.utk.edu/

Dr Donald Bouldin

https://www.eecs.utk.edu/people/faculty/dbouldin/

During my efforts to lead the university program, Dr Bouldin was a constant source of advice and counsel. His approach was more pragmatic as he had industry experience and that of consulting. He became a key player in my efforts to help define Mosis Tiny Chip with NSF

https://www.mosis.com/pages/products/mep/mep-about#instruct

National Science Foundation

https://www.nsf.gov

Dr Bernard Chern

https://www.nsf.gov/about/history/nsf0050/manufacturing/history.htm

Dr Chern was a leader of Mosis program in the 80s with whom I presented the idea which then morfed into Tiny Chip.

Preface

In life our actions are formed as a sum of our experience as people. This is fundamentally what sets us apart and makes us individuals. It also defines the guidebook for our journey. In this book I am not to suggest that this is the Rosetta stone, guide to making a computer actually think like us or create R2D2, Terminator or a Genie. On the contrary, this volume is dedicated to simple observations gathered, borrowed or collected over a lifetime of immersion in the world of Math, Science, computing and Anthropological study.

To prove validity of some notions within I will offer mathematical support. As time is not unlimited, I will leave proof to the reader in other areas where I have sown seeds of thought. I grew up on a farm, studies the effect of music and poetry on people. Find problem solving a blast and experiment creation a rush. Essential ideas herein are meant to be within the grasp of anyone with interest. Heretofore, consider this a book of suggested filters, lenses through which computing can improve in effectiveness. I would never suggest that computing as we know it is deeply flawed but at times confused. Some will say to make a mistake is human, to really screw things up, takes a computer. My goal here is to suggest a few simple ways in which computers can be adapted to make them far more intuitive and less frustrating as a tool to get something done.

The lawyer in me must add this note. Ideas within this book are mine and mine alone. I am not representing any company, institution or government in my opinions. Consider this a set of hypotheses organized in chapters with the intention of stirring the pot, exciting an effort, taking us outside the box, to observe and create.

A Picture Is Worth One Thousand Words, Choose Them Wisely, Maybe

To set the stage for maybe requires and understanding of yes and no in computing. Long ago someone began tabulating with marks, notches, beads, bundles, stacks, etc. Someone else noted the polar nature of magnets. In general most people find comfort in thinking in concrete terms as though things are cast in stone. Black or white, yes or no, on or off seem to be the preferred choices of answers to questions. This seems to be rooted in the concept that seldom does anyone want to "do the math".

Thus, when another person discovered the PN junction and subsequent transistor, amazing things grew. Transistor radio was a real thing. One could carry a small, lightweight device to extract from the air music, stories, events. All this grew from

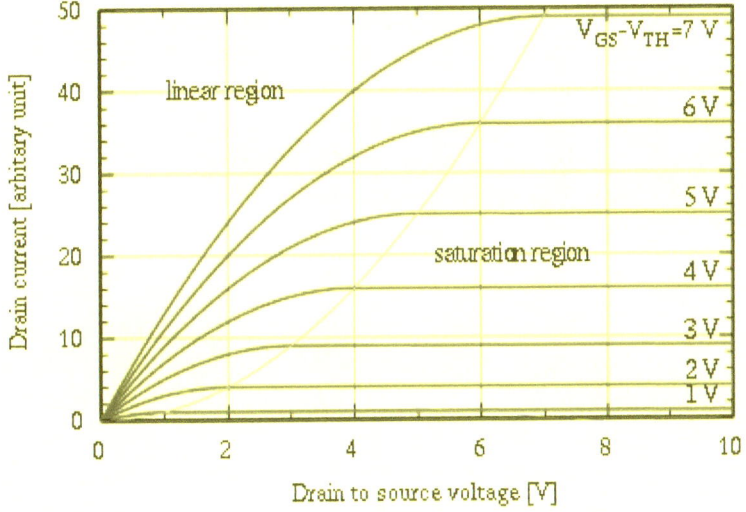

some really simple facts:

Take an effective insulating material which happens to grow in an anal retentive fashion, ie silicon. It happens to arrange itself in a crystalline fashion such that all atoms line up in 3 dimensions. A shining example is a diamond. This material catches the eye, glimmers in the sun. With this arrangement of atoms there are regular, predictable spaces in which other atoms can diffuse carrying loose charges (electrons). When acted upon by an electric field, the electrons tend to flow like a dry creek bed during a heavy rain. The amount of flow depends on the applied force. See classic

Figure 1 from https://en.wikipedia.org/wiki/Current%E2%80%93v oltage_characteristic

From this figure one can see that the coarse filter through which modern computing tends to view a transistor is represented by the red curve above and restrict to a small band of operations in the y axis. In essence, an amazing 3 terminal device with complex characteristics has been dumbed down to act like a mechanical switch for on or off. The devil herein is in the details. The image suggests that with 0 applied source voltage a transistor does not have any current or movement of electrons, exactly none. In reality, unless you unplug a battery from electronics, stuff still happens when a transistor is told to be "off". Common terms for this tend to be leakage current. More on this phenomena in another installment.

Suffice it to say, to really get your head around how a transistor works, usually requires lots of Math and tireless attention to science. This study is commonly known as Electrical Engineering. Most people would prefer to learn how to use computers to get something done, not the guts of how it works much the same as getting in a car, turning the key and pointing the car. Along the lines of taking this complex device and applying it to making computers there was a ground breaking book written in 1978 by Carver Mead and Lynn Conway called Introduction to VLSI Systems. This book and associated undergraduate courses revolutionized the speed with which one can create and manufacture widgets that compute.

Despite the success of the aforementioned, I will divulge here a few transistor facts which we tend to summarily ignore today that could be important.

1) There is no real off without complete disconnect (look critically on the software parallels of laptops, phones, tablets these days)

2) Linear region offers a high precision multiply (look at audiophiles and pure tone replication)

3) There are shades of on (inherent multi-value states in each transistor)

Computing today, in large part, relies on the premise that all tasks can be reduced to a series of questions to which the response is yes or no, definitively. With this simplification people traveled to the moon, drove cars, flew satellites, searched vast databases to good effect. That said, there is a simple change to existing methods that can greatly impact the effectiveness of computing and that is the inclusion of maybe in the vernacular.

At first glance you may say maybe sounds too fuzzy to get your head around. I will begin by acknowledgement of fuzzy logic.

Figure 2 from https://en.wikipedia.org/wiki/Fuzzy_logic#Overview

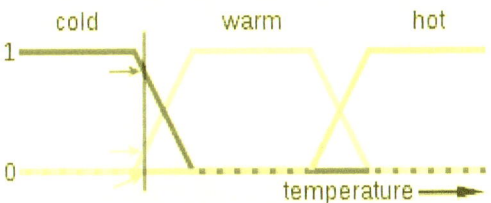

I will next say that my suggested method of maybe is not reliant on fuzzy logic in any direct way. One could suggest that my definition is included in the superset called fuzzy logic but my approach differs fundamentally. The distinction I draw is that fuzzy logic is more like a sum of yes or no results over the axis of time. Maybe is really the encapsulate of possible or good idea in companion with certain(1) or never(0).

To illustrate my thought I have a few examples of common instances of Maybe.

Notes On Big Picture

Lovely Rita Meter Maid

In your home everyone has meters running that merely tabulate use of things. Examples include gas meter and water meter. These operate at the core on mechanical basis of spinning gears that serve to dutifully record the flow of resources into, and in the case of electrical energy, out of, your home. None of these methods of measurement require a computer to do their task of monitoring. These days, they are connected to electronics which can sample usage and alert utilities to problems or in the case of energy management, be used to load balance electrical demand and avoid widespread blackouts. This is the topic of another day on smart meters.

With all this infrastructure, one would think the days of meter reading would be a thing of the past. To this day, the electronic surveillance of utility use is augmented by "boots on the ground", aka meter readers. Herein lies my first example of maybe. You agree to pay monthly for service but the exact time the meter is read is time averaged in time to be monthy. In this way, each month the meter is definitely read within a few days of expected and the result is you have largely predictable cost. However, if I asked anyone to predict the moment in time the meter person would arrive, it is unlikely the answer would be yes or no, this is a case for maybe.

If I told you there is a promotion which offers a 10% discount but requires you to accept the risk that you will have your electricity shut off exactly during the hottest part of the day, you would have to way is it worth it? If you think it is worth it, you internalize maybe as your answer. This is illustrating calculated risk taking.

Herein I want to pull out my first two filters of computing for optimization.

Filter 1 – Mindless Meters

Kick off background tasks for counting, accumulation, sampling which tabulate in memory or disk but minimize the memory footprint by strict rules of memory budgeting (aka mindless meters)

Filter 2 – Measure Precision

Sample meters output on rounds based on required precision of units and allowable measurement error.

If I return to the analogy that computers are constrained to decisions based on a binary decision tree of yes or no, adoption of maybe implies the ability to skip a major series of questions or tests based on gut feeling (cursory review of past history) Then, if things go badly, take note (been there, tried that). If this path resulted in desired outcome, weigh it as continued favorite path.

One implementation of maybe is to form a new logic function based on the harvest of linear region in a transistor. A likely response may be "impossible", to which Dr Heinz would demand proof. I would not envision complete replacement of our current logic families but rather incorporation in mission or performance critical decision trees and it should be augmented with error detection/correction.

Another, equally viable approach is to add a software overlay. In this scheme, final decision points could factor in a maybe bit as an or function or a leg of a mux in the control or datapath. When set, this bit, in effect tells the computer go for it. This could be the reset state or a more conservative option of do the work.

Notes on Meters

Can't Does Not Compute

To move beyond current methods of computing we need to believe that we can. This was affirmed by my Grandmother as follows: "You can say you dont want to, say you choose not to, even say Hell No but if you ever say you can't, you lie". While I am not suggesting these new models will be trivial to implement or predict the actual day of the first implementation. I will say, it will.

Notes on Can

Constraints Are Your Friend

My mother always said that "necessity is the mother of invention". I suspect this was an adage helpful to a person who saw two "world wars" and the great depression where no one lived with an abundance of things, food, etc. This comes to my next filter.

Notes On Constraints

Filter 3 – Break It Into Tiny Fragments of Memory

Slice your problem into tiny pieces and cast out several tiny tasks. Tiny in the memory footprint with very finite needs and crisp results of yes, no or maybe. Send out at least as many monitor steps in inspect progress. These monitors can flush a task from memory or cut it short if wheel spinning is detected. On a compute server tightly constrain active jobs to allow yours to assume highest priority. Demand that all programs release memory when done and allocate in a fashion coincident with linked list allocation. Make use of checkpoint/restart functions to store relevant threads in statics storage for recall.

Notes On Fragmentation

Do The Math or Simply Graph It

In a world of clouds filled with sensors where your location and communications are not completely private, you need to make conscious choices to apply a filter under your control. Despite the notion that Big Brother is all knowing and all powerful, in order for someone to receive data you need to permit transmission.

Notes On Do The Math

Then, Again, Set It To Music

Think about the inherent reaction each of us has to music. Realize that compositions of music transcend language. This is why one can enjoy compositions of music, classic or otherwise where the native tongue is immaterial. My hypotheses on this is with or without lyrics, music is the carrier wave on which information rides.

You may now saw, huh? Well, here is the deal. When people listen to music, most often, they use their audio system (cochlea) as a filter. For a while, in 1979, I did some research on trying to build an artificial cochlea and contemporary though was to build receiver of audio signals, transform the range of frequencies to another range that a person can hear which conceptually is called compression amplification. While in undergraduate school I built and validated a thick film hybrid microelectronic device that worked well. To this day, the state of that art in this realm has not fundamentally moved from this concept though my version would.

Notes On Notes

Cochlea Is A Simple Machine

To understand my next computing filter requires a simple overview of a simple machine, the cochlea. In terms of math, this "device" is an array of pure tone filters (hairs) which are passive devices. They naturally resonate in the presence of a tone of matching frequency (wavelength). Much like a tuning fork, can wade through white noise to stand up and shout when a pure tone is noticed. Even to the untrained ear, most people can detect a sour note.

So here is the basic math tied to this machine

cochlea = Σ (pure tones detected)

Where pure tones detected(i) = raw tone(i) * tone mask(i)

This basic arrangement allows appreciation of such "simple" things as harmonies. These are perfectly synchronized arrangements of pure tones where the tones mixed are either identical or an integral multiple of one another in pitch. In some sense, this is another example of implicit math, as each hair has finite and precise length. Hairs that are integral multiples of one another will form a logical and function in the brain.

Notes On Simple Machine

Filter 4, Tuning Forks

This is a concept I developed in observations of building silicon from 2um to present day. Metal lines, regardless of dimensions, act physically and electrically similar. Thus, it is viable to build an analog array of tuning forks which vibrate at resonant frequency and are completely passive (take no excitation energy other than auditory.

Notes on Cochlea

Notes on Tuning Forks

Simple Example Of Using Simple Machine To Do "Complex Math"

Given one has built the aforementioned low power machine. Matters of cataloging genres of music can be relegated to sampling simple bursts of the opening phrases of a song. In addition, this method can crisply identify source band and likely era created, album in which song was introduced. If you were to try this at home, in math derivations, it would take pages of tiresome calculations.

Let this serve as the first installment on a new series to support challenging the norm of computing and suggesting new applications of electronics/software to ease the pain of computing. While this is a brief introduction of topics that merely skims the surface, I can talk at length on any of these elements and would be delighted to do so.

Feel free to ping me on Linked In (https://www.linkedin.com/in/edwardmseymour?authType=NAME_SEARCH&authToken=oYIp&locale=en_US&trk=tyah&trkInfo=clickedVertical%3Amynetwork%2CclickedEntityId%3A138532144%2CauthType%3ANAME_SEARCH%2Cidx%3A1-1-1%2CtarId%3A1473778056131%2Ctas%3Aed) or Gmail (edwardmseymour@gmail.com).

Notes On Start The Dialog

www.ingramcontent.com/pod-product-compliance
Lightning Source LLC
Chambersburg PA
CBHW021852170526
45157CB00006B/2419